Electricity

Karen Bryant-Mole

RIGBY
INTERACTIVE
LIBRARY

This edition © 1997 Rigby Education
Published by Rigby Interactive Library,
an imprint of Rigby Education,
division of Reed Elsevier, Inc.
500 Coventry Lane
Crystal Lake, IL 60014

© BryantMole Books 1996

Printed in Hong Kong

00 99 98 97 96
10 9 8 7 6 5 4 3 2 1

Library of Congress Cataloging-in-Publication Data
Bryant-Mole, Karen.
     Electricity / Karen Bryant-Mole.
       p.   cm. — (Science all around me)
     Includes bibliographical references and index.
     Summary: Explains the basic principles of electricity through looking at everyday experiences and direct observation.
     ISBN 1-57572-109-0
     1. Electric power—Juvenile literature. 2. Electric apparatus and appliances—Juvenile literature.
[1. Electricity. 2. Electric apparatus and appliances.] I. Title. II. Series.
TK148. B79   1996
621.31—dc20   96-22977
                                                               CIP
                                                               AC
Cover designed by Lisa Buckley
Interior designed by Jean Wheeler
Commissioned photography by Zul Mukhida

Acknowledgments
The publisher would like to thank the following for permission to reproduce photographs. Chapel Studios, p. 16; Eye Ubiquitous, p. 6; National Power, p. 8; Positive Images, pp. 9, 12, 14; Tony Stone Images (Paul Redman), p. 4 (Mary Kate Denny), p. 10; Zefa, pp. 18, 20, 22.
Every effort had been made to contact copyright holders of any material reproduced in this book.
Any omissions will be rectified in subsequent printings if notice is given to the publisher.

> **Note to the Reader**
>
> Some words in this book are printed in **bold type.** This indicates that the word is listed in the glossary on page 24. This glossary gives a brief explanation of words that may be new to you and tells you the page on which each word first appears.
>
> Answers to the questions on pages 4, 6, and 14 appear on page 24.

Visit Rigby's Education Station® on the World Wide Web at http://www.rigby.com

Contents

What Is Electricity?

Electricity is a form of **energy.**

Energy can be difficult to understand because it cannot usually be seen.

One way to describe energy would be to say that it makes things work.

Lots of the things that we use every day need electricity to make them work.

? Which objects in this picture need electricity to work?

See for Yourself . . .

We often use electrical things
in a particular room in the house.

For example, a toaster is usually
found in a kitchen.

Draw a big picture of
a house. Cut out from
magazines some
pictures of things
that need electricity.
Glue the pictures in
the rooms where they
will be used.

Static Electricity

During a thunderstorm, tiny water droplets in the clouds rub together and become charged with **static electricity.**

The electricity in the storm clouds then moves between the clouds or down to the earth.

It makes a very big spark.

? *What is this big spark called?*

See for Yourself . . .

Pour some salt on a paper plate. Rub a plastic spoon very hard on your shirt. Hold the spoon just above the salt. What happens?

Static electricity makes the salt jump up and cling to the bottom of the spoon.

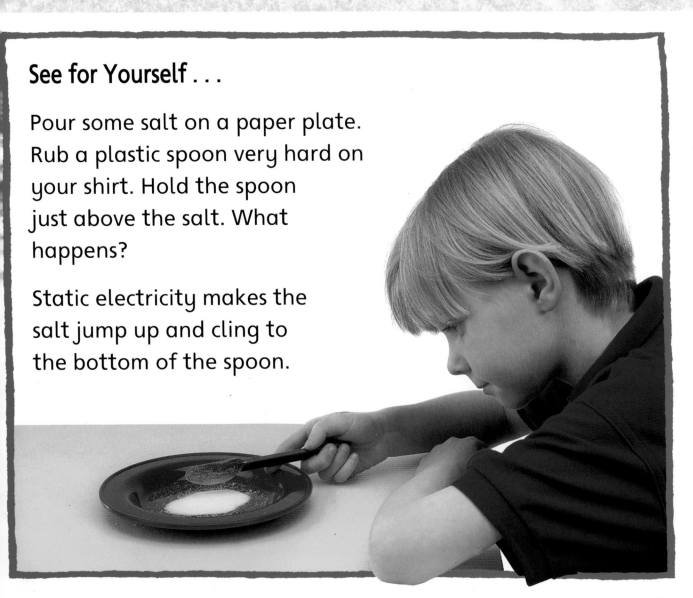

Manufactured Electricity

Electricity can be made, or manufactured, in a special building called a *power station.*

Electricity is moved along power lines to homes and businesses.

Electricity that moves from one place to another is called *current electricity.*

? *Can you see some wide towers? They are called* **cooling towers.**

See for Yourself . . .

As electricity enters our homes, it passes through a **meter** that measures the amount we use.

Ask an adult to show you the electricity meter outside your home.

Look for a dial spinning.

The more electricity your family uses, the faster the dial spins.

Using Electricity

Electricity is one form of **energy**. It can be changed into other forms of energy.

Electricity can be changed into heat energy by a toaster, light energy by a light bulb, or sound energy by a radio.

(i) *Electricity can also be used to turn a **motor**. Dishwashers and vacuum cleaners have motors.*

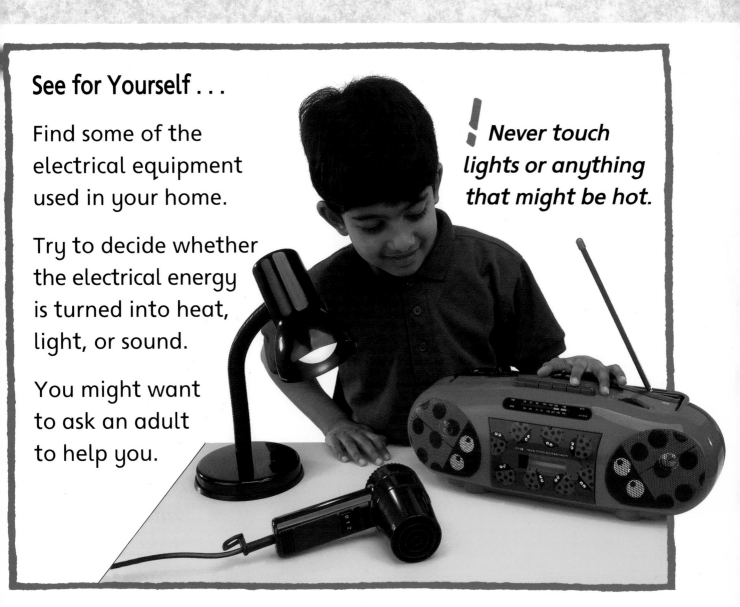

See for Yourself . . .

Find some of the electrical equipment used in your home.

Try to decide whether the electrical energy is turned into heat, light, or sound.

You might want to ask an adult to help you.

! *Never touch lights or anything that might be hot.*

Dangers of Electricity

The electricity in your home is very useful, but it can also be very dangerous. It is powerful enough to kill you.

Pylons that carry power lines across the countryside have warning signs on them. The electricity in the power lines is very dangerous.

Never play with plugs, light bulbs, sockets, or wires.

See for Yourself . . .

Signs, like the one on the pylon, warn people about the dangers of electricity.

Design your own electricity warning sign. Make sure your sign is brightly colored so that it can be seen easily.

Watch for other signs that warn of the dangers of electricity.

Batteries

This personal stereo is powered by batteries.

Batteries have chemicals inside them that can change and make electricity.

When all the chemicals have been used up, the battery stops making electricity.

? What other objects are powered by batteries?

14

See for Yourself . . .

Put a battery in a toy that uses one battery. Turn on the toy. Does it work? Why or why not?

If the toy does not work when you turn it on, it means that the chemicals in the battery have been used up.

Batteries like this are safe to use because their electricity is much weaker than the electricity flowing around our homes.

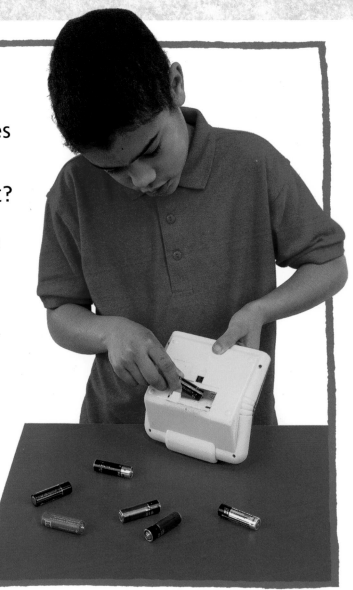

Circuits

The bulbs and wires in these Christmas tree lights form a *circuit*. A circuit is a loop around which electricity can flow.

If there is a break in the circuit, the electricity is not able to flow all the way around it.

(i) *If one of these bulbs stops working, all the lights might go out.*

See for Yourself . . .

Try using a battery, a bulb in a holder, two pieces of wire, and some tape to make a circuit.

Does the bulb light up? Why or why not? Electricity cannot yet flow all the way around the circuit.

If you join the end of the wire to the battery, it will make a loop and the bulb will light up.

Switches

When you want to watch TV or turn on a light, you press a button or flick a switch.

Switches are used to **complete** or to break a circuit.

Switches turn equipment on or off.

?*How is the electrical equipment in your home turned on and off?*

See for Yourself . . .

You can make a circuit with a switch. Join three pieces of wire, two thumbtacks, a paper clip, and some **balsa wood**.

You can use the paper clip switch to complete or break the circuit.

This makes the bulb flash on and off.

Conductors

Materials that let electricity flow through them easily are called *conductors.*

Most metals are good conductors.

Trolleys use metal wires to conduct electricity from overhead wires to the trolley.

Our bodies conduct electricity, too. If electricity passes through someone's body, it gives an electric shock.

! ***Never play with electricity.
An electric shock can kill.***

See for Yourself . . .

Use your circuit to see how a conductor works.

Fold a piece of foil into a strip. What happens when you tape the ends of the wires to the ends of the strip? The bulb lights up because foil conducts electricity.

Insulators

Materials that do not let electricity flow through them are called *insulators*.

Most plastics are insulators.

This iron is covered in a plastic case to protect the person using it.

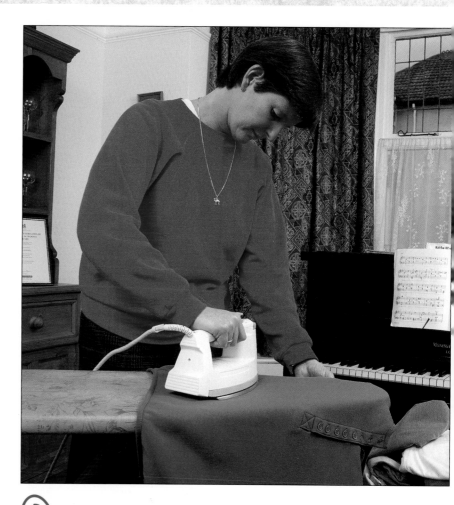

ⓘ *The wires that carry electricity in homes are usually covered in plastic, too.*

See for Yourself . . .

Try to find out which materials are insulators and which ones are conductors.

Find some objects around the house and use tape to connect them into your circuit.

If the bulb lights up, the material that the object is made from is a conductor. If not, the material is an insulator.

Glossary

balsa wood   a light type of wood   19

complete   to finish, make whole   18

cooling towers   places where water that has been used in the power station is cooled down   8

energy   source of power   4

meter   a machine that measures something   9

motor   an engine that makes things work   10

pylons   tall, metal towers used to carry power lines   12

sockets   the holes into which a plug is pushed   12

static electricity   electricity that stays in one place   6

Index

Further Readings

Berger, Melvin. *All About Electricity*. Scholastic, 1995.

Cooper, Jason. *Electricity*. Rourke Corporation, 1992.

Peacock, Graham. *Electricity*. Thomson Learning, 1994.

Answers

p. 4, clock, dryer, iron, lamp, washing machine; p. 6, lightning; p. 14, toys, radios, flashlights.